Push [illegible] Pull Faces

by Allison K. Lim

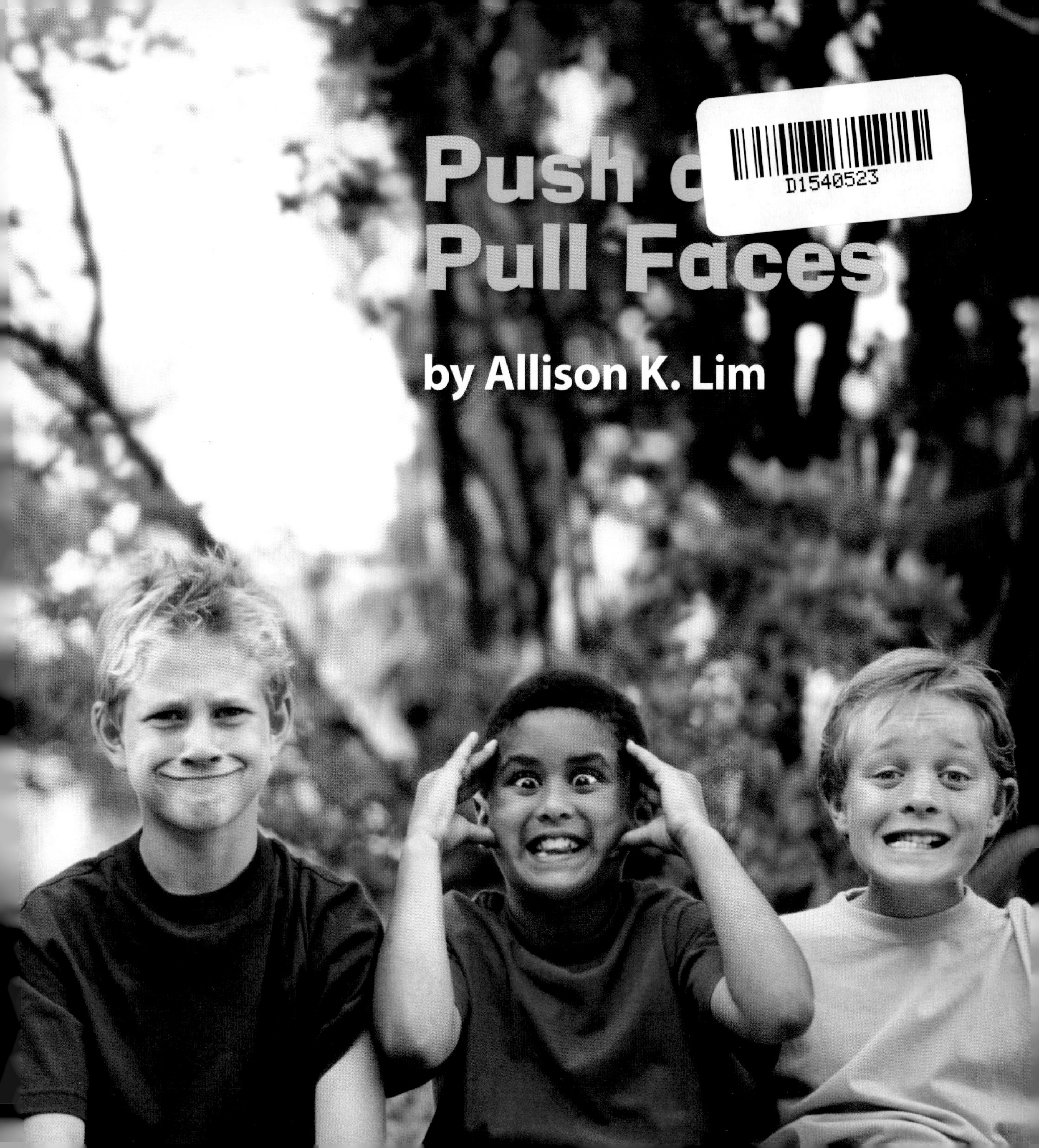

Push

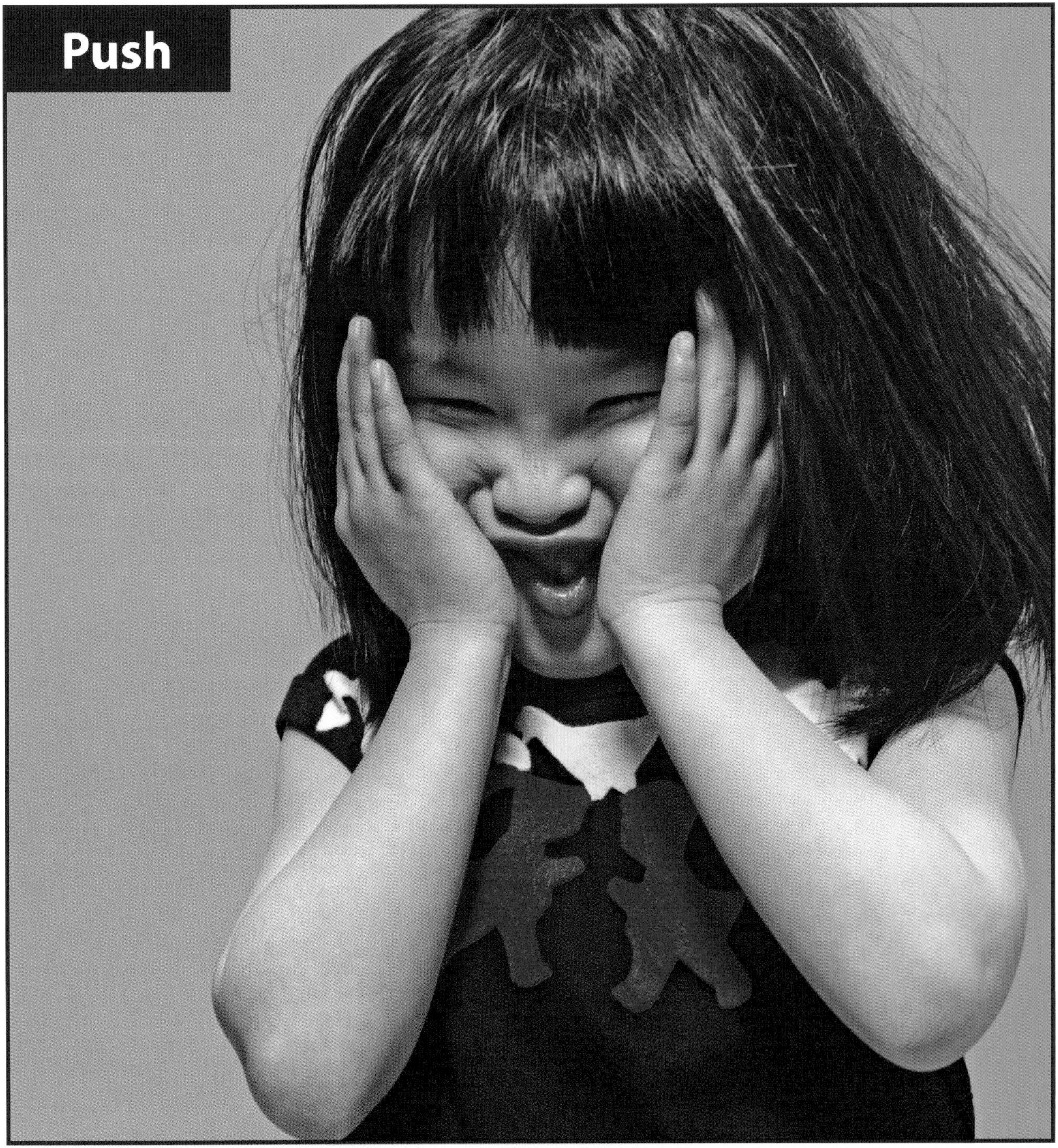

Pull

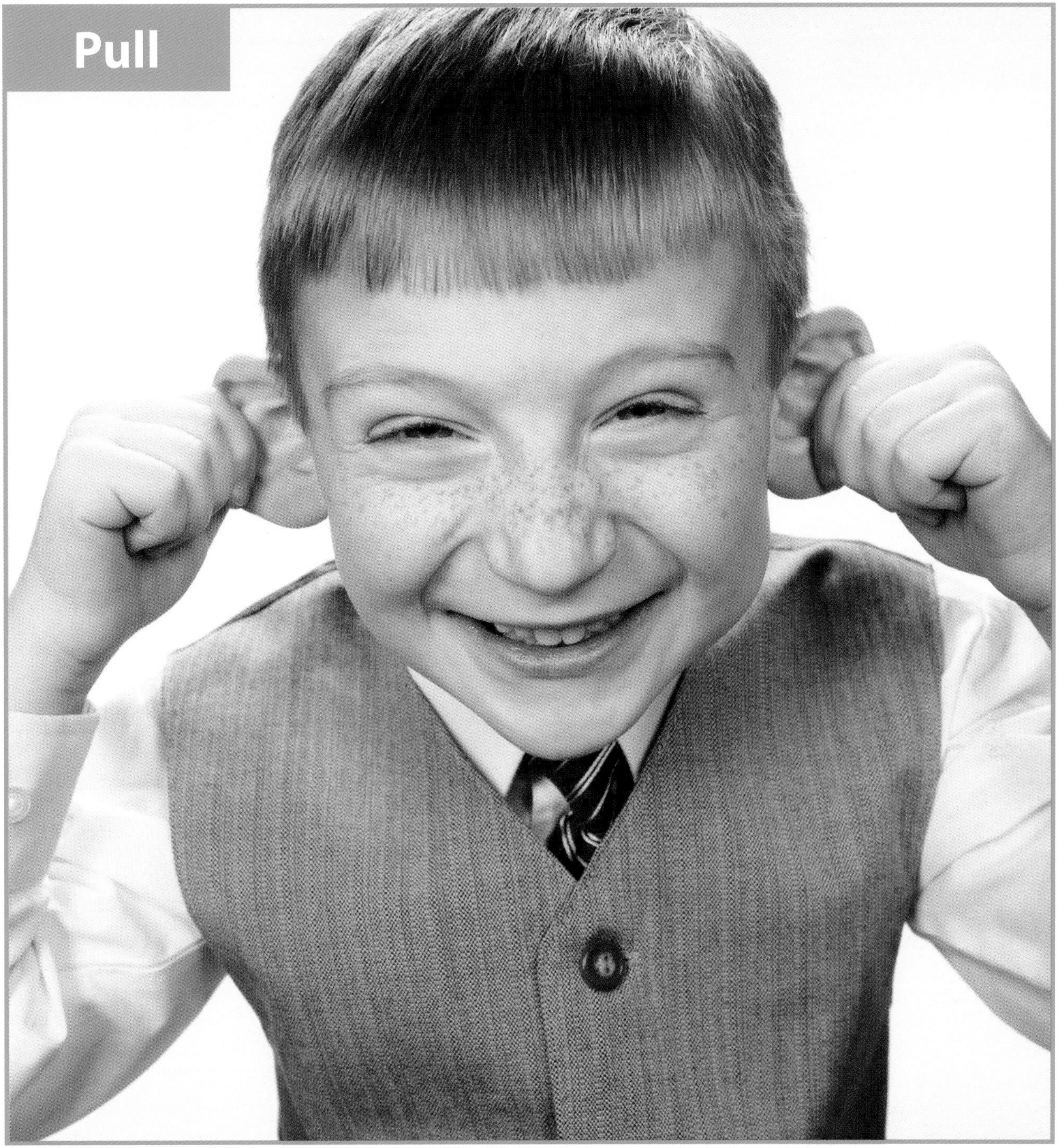

Push

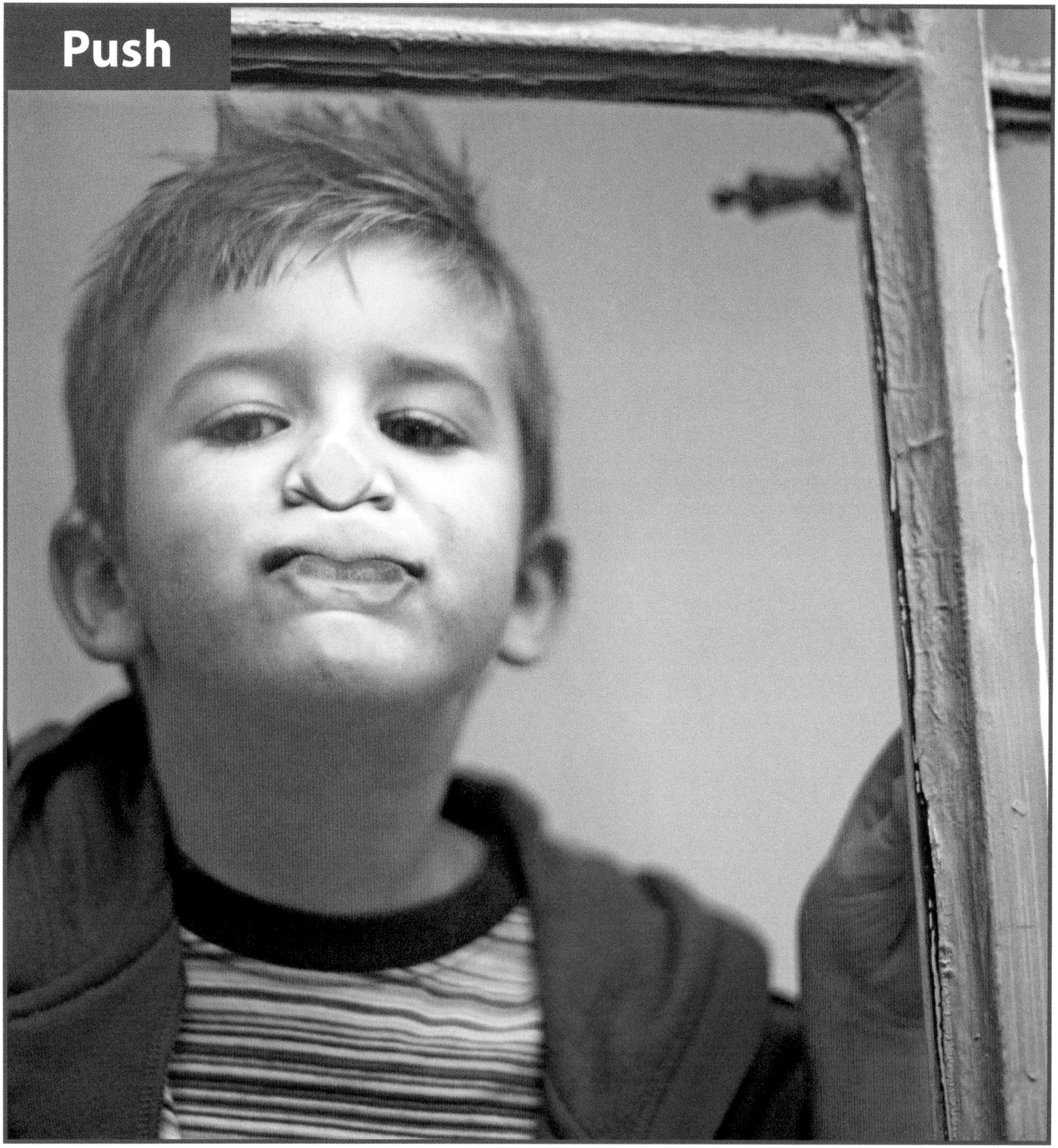

Pull

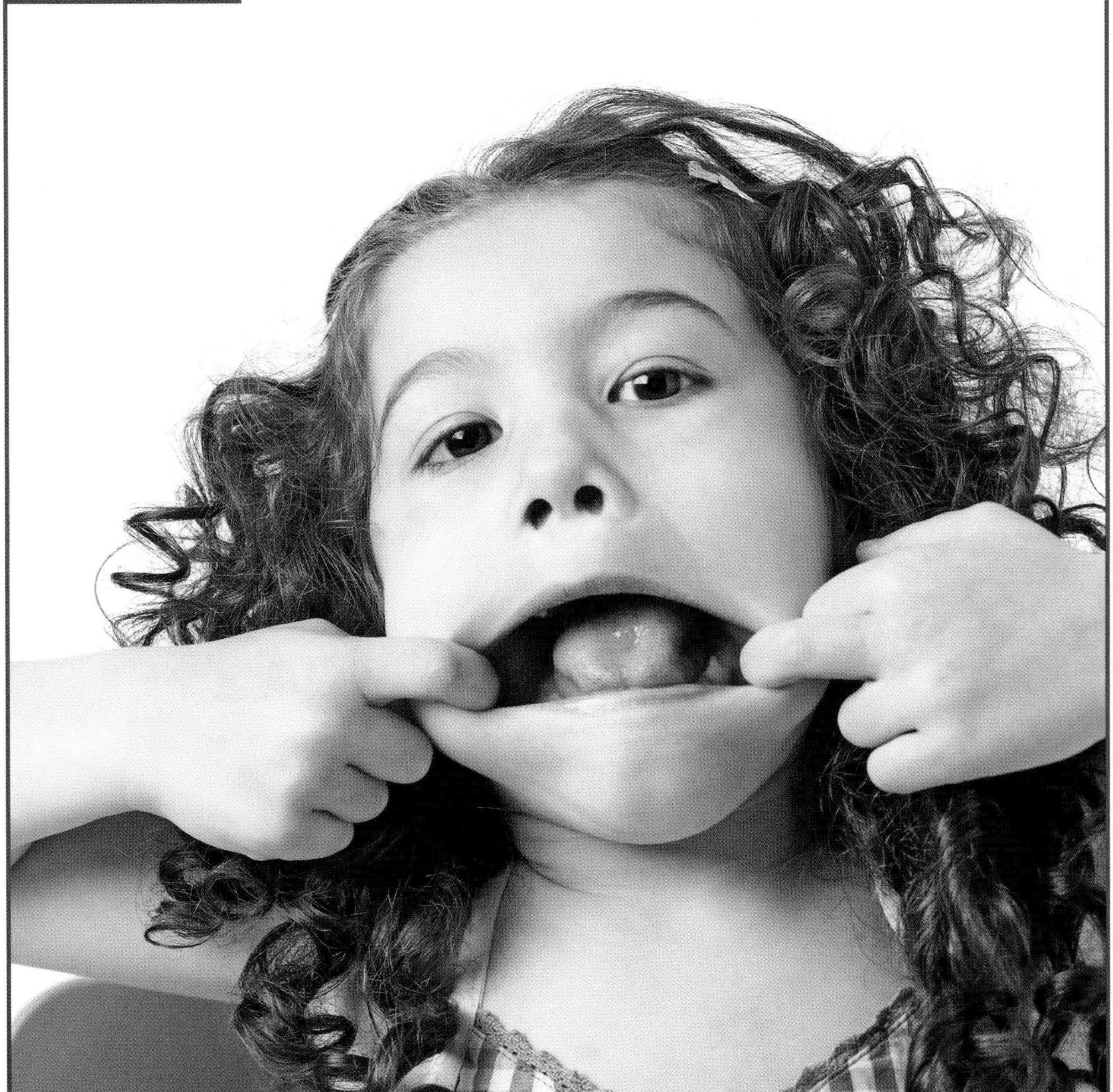

Push

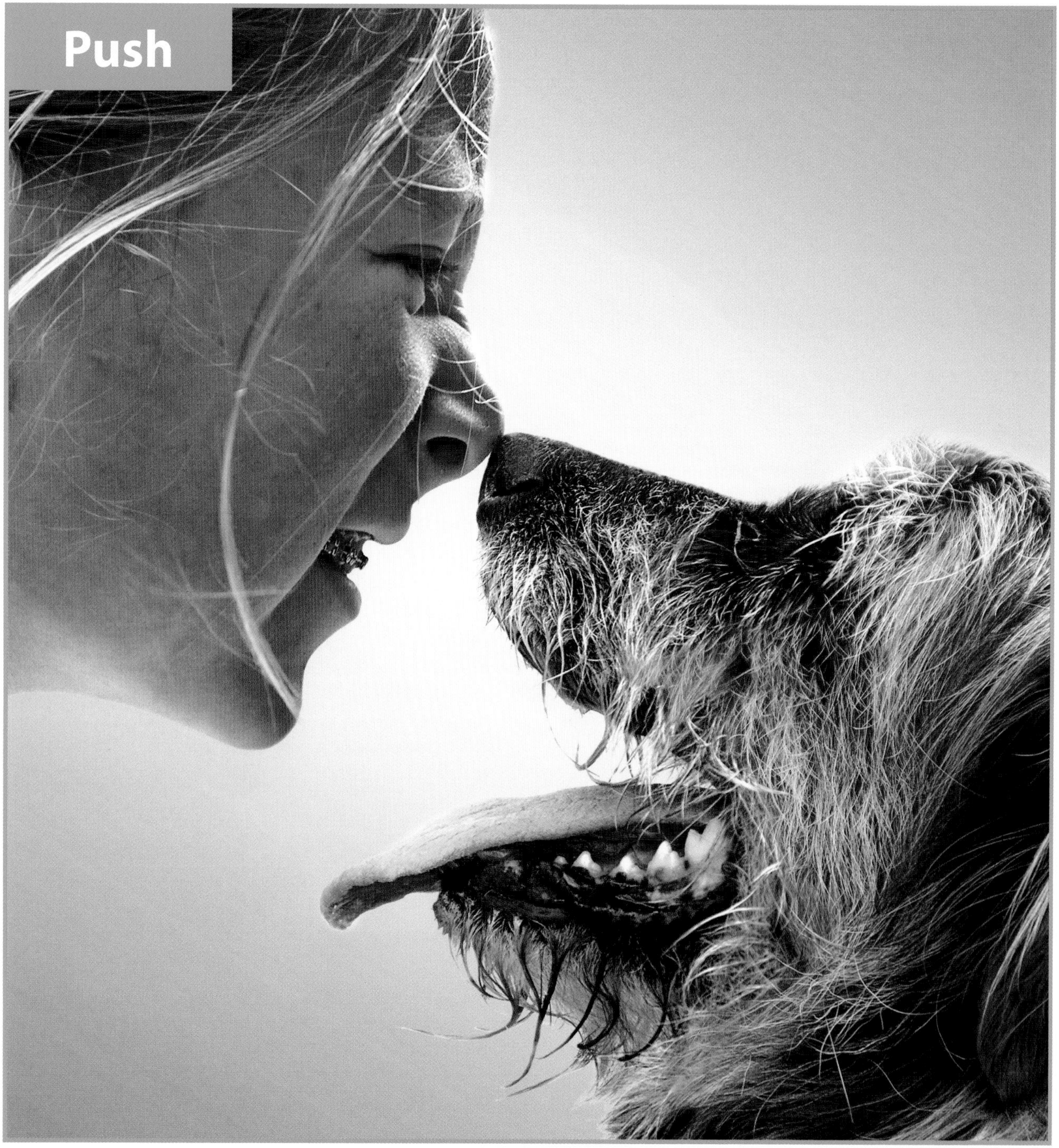

Pull

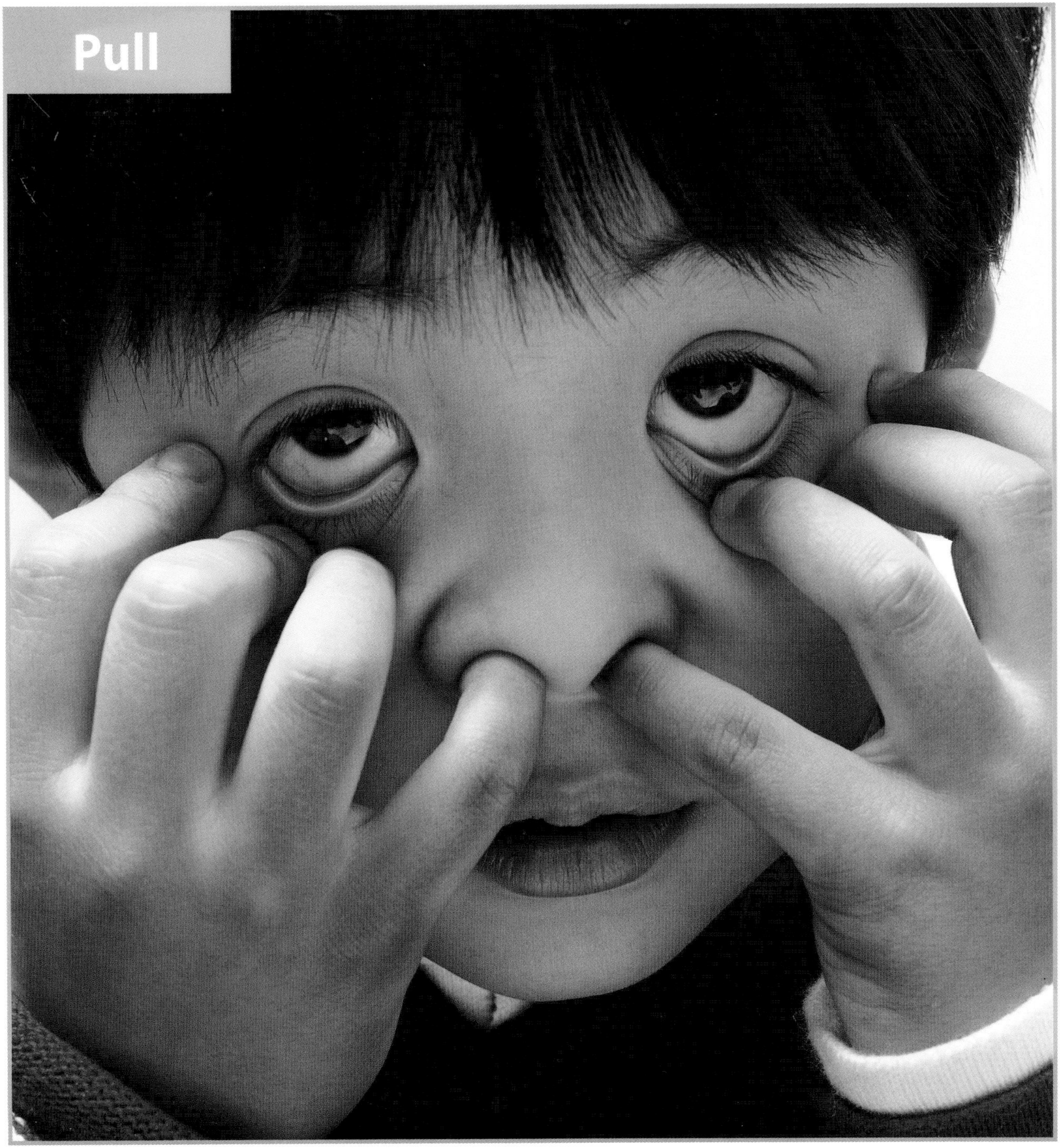

Push

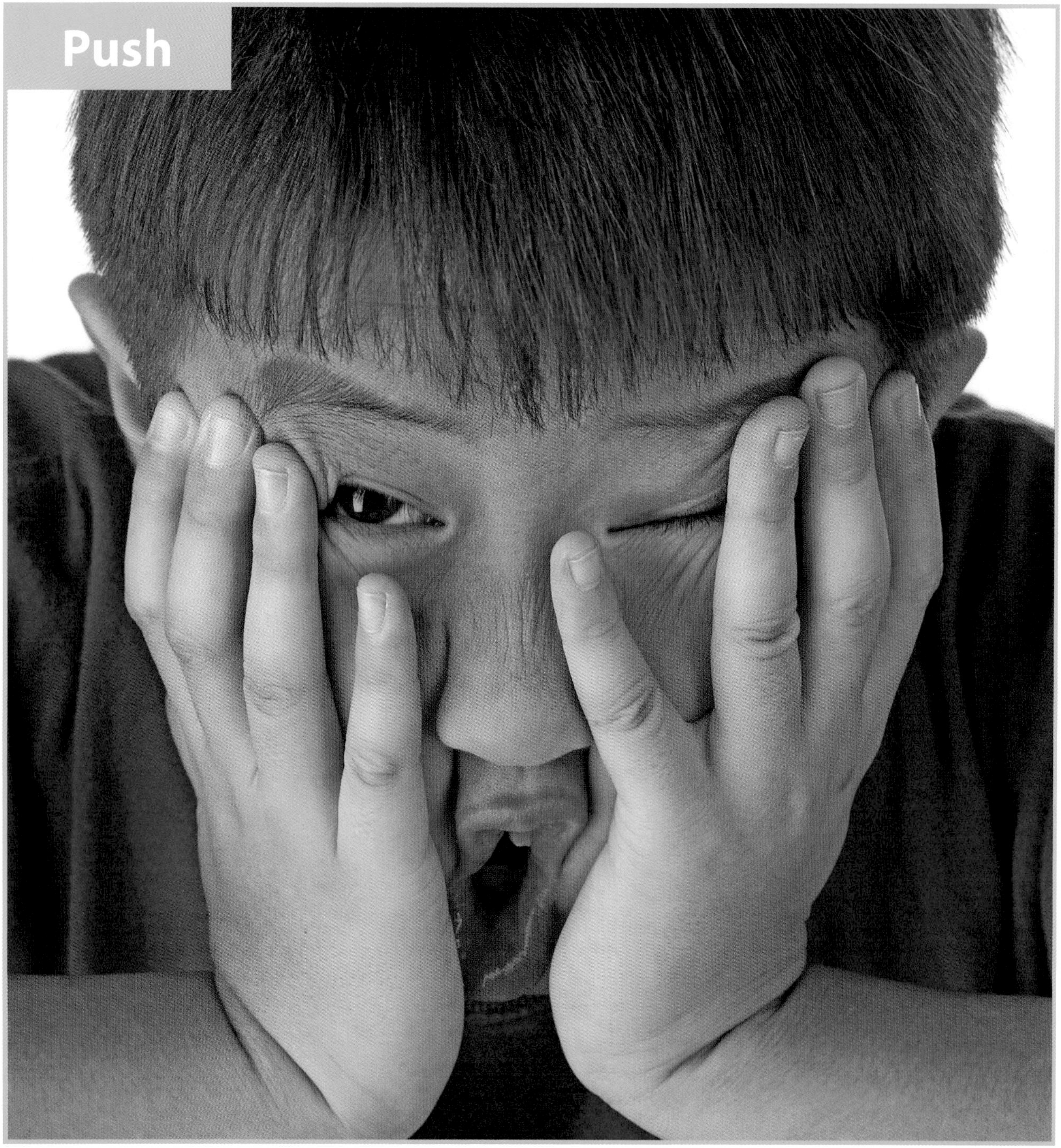

Pull

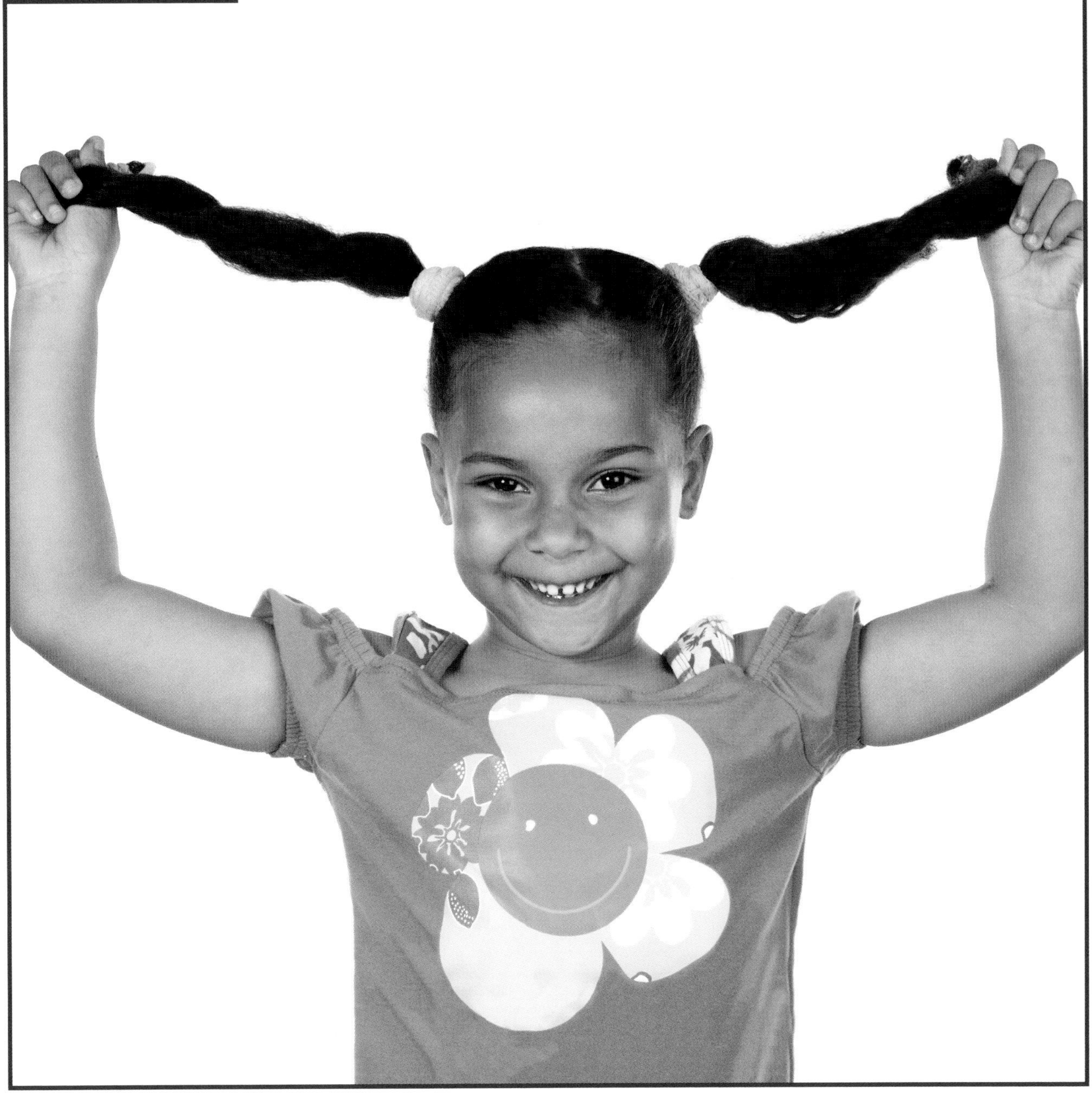

Push

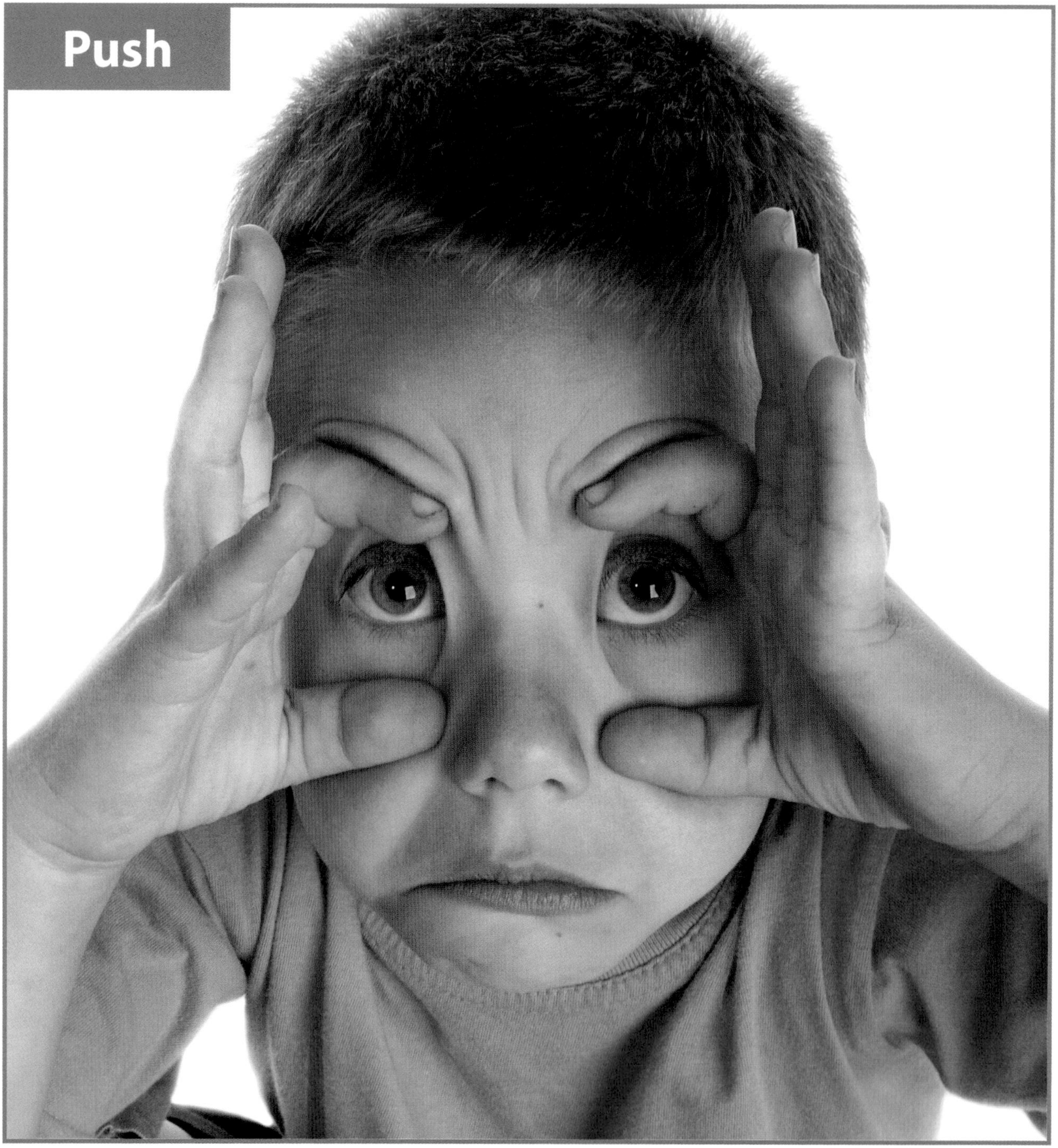

Pull

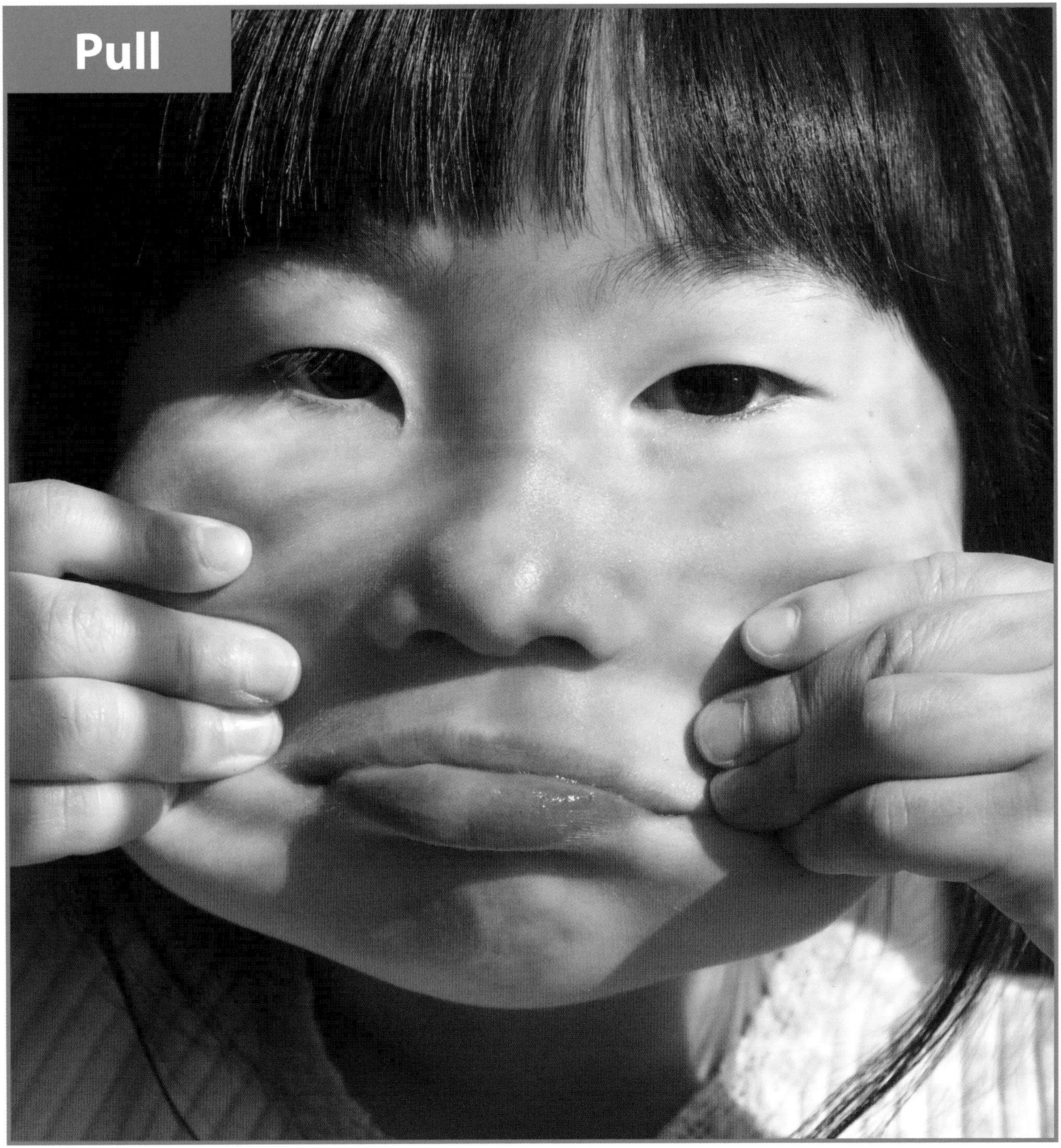

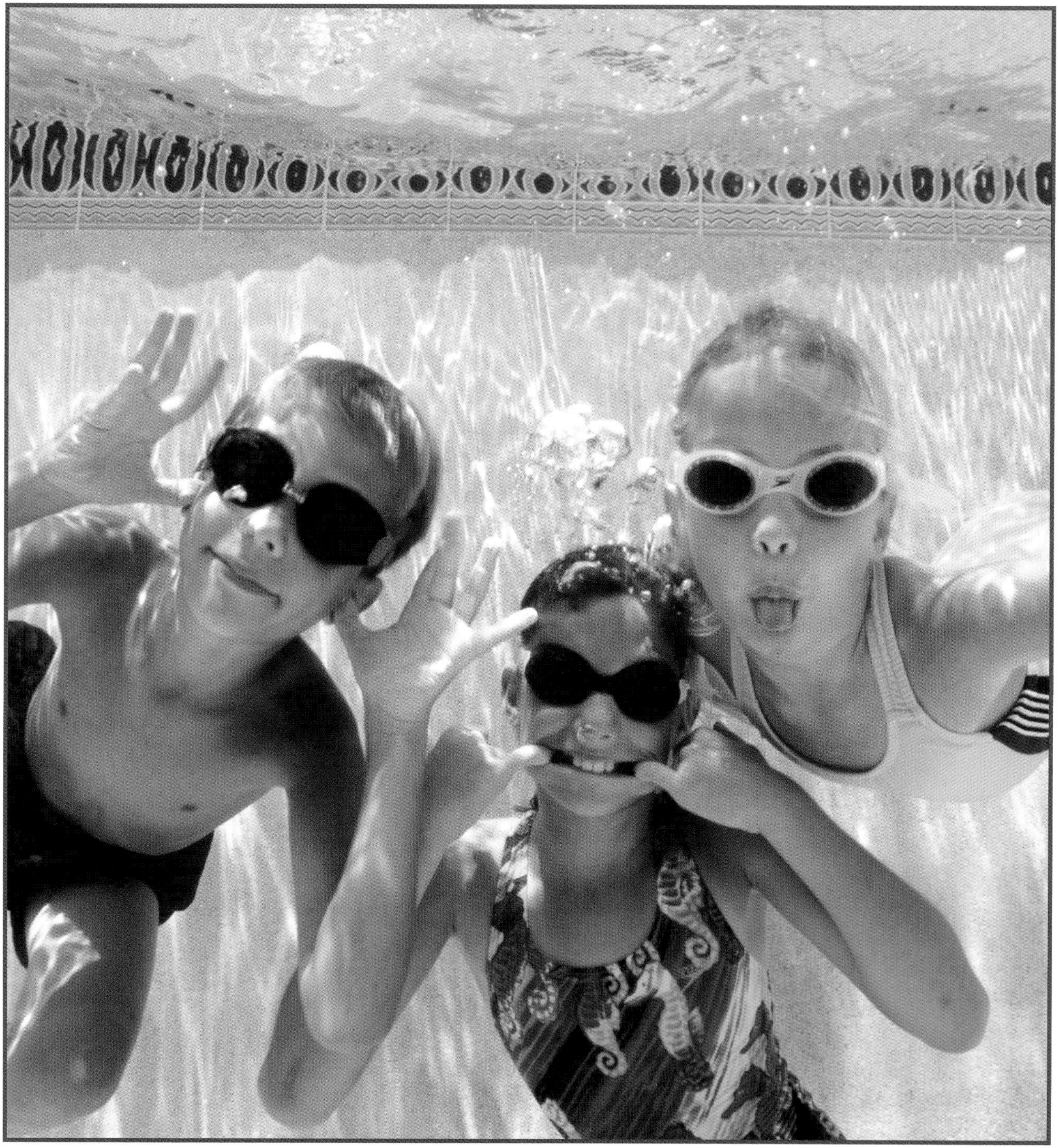